Editors' Handbook

1988

AMERICAN SOCIETY OF AGRONOMY
CROP SCIENCE SOCIETY OF AMERICA
SOIL SCIENCE SOCIETY OF AMERICA
677 South Segoe Road • Madison WI • 53711 • (608) 273-8080

ACKNOWLEDGMENT

This *Editors' Handbook* was developed by the Tri-Society Editorial Policy Coordinating Committee, consisting of D.R. Buxton, G.H. Heichel, E.L. Horner, C.W. Stuber, J.J. Mortvedt, and D.E. Kissel. A number of society members and editors, including R.G. Hoeft, R.R. Johnson, T.J. Logan, and G.A. Peterson, assisted. Also involved in the production was a headquarters office committee consisting of D.M. Kral, R.C. Dinauer, W.R. Luellen, S.H. Mickelson, and Susan Ernst, with contributions from all members of the headquarters editorial staff. Barbara S. Littlewood, a professional indexer, contributed to the section on index preparation. William R. Luellen served as managing editor and Patricia Jeffson designed the cover.

American Society of Agronomy, Inc.
Crop Science Society of America, Inc.
Soil Science Society of America, Inc.
677 South Segoe Road, Madison, Wisconsin 53711 USA

Library of Congress Cataloging in Publication Data

American Society of Agronomy.
Editors' handbook.

1. Agricultural literature—Editing—Handbooks, manuals, etc. I. Crop Science Society of America. II. Soil Science Society of America. III. Title.
S494.5.A96A44 1988 808′.06663 88-16676
ISBN 0-89118-095-8

Printed in the United States of America

Contents

Chapter 1. Introduction

The American Society of Agronomy, Crop Science Society of America, and Soil Science Society of America (hereinafter, referred to as the three societies) publish information concerning all aspects of the sciences they represent. Their extensive publishing activities offer opportunities for professionals to communicate findings to colleagues and to larger, more general audiences.

The *Publications Handbook and Style Manual*, 1988 serves as a guide for authors in preparing manuscripts and other material submitted for publication by the societies and for editors in evaluating the manuscripts and material. This *Editors' Handbook* is intended to serve as a further guide to editors in carrying out their responsibilities.

The three societies publish six journals, *Agronomy Journal*, *Crop Science*, *Soil Science Society of America Journal*, *Journal of Production Agriculture*, *Journal of Environmental Quality*, and *Journal of Agronomic Education*. One or more of the societies also publish three book series—Agronomy Monographs, the SSSA Book Series and Special Publications as well as other miscellaneous books, papers, booklets, and pamphlets. Two other titles, *Agronomy News* and *Agronomy Abstracts*, are edited and published by headquarters staff and will not be discussed in this handbook.

The general editorial policies and practices of the three societies are determined by the respective boards of directors, whose members are elected.

The boards of directors of the three societies have delegated the responsibilities for review of manuscripts and the production of publications to the editors-in-chief, editors, editorial boards and committees, and to a full-time staff at the society headquarters office.

Day-to-day procedures and practices have evolved through the years and are still changing. Much of the paperwork and all of the printing, production, and distribution of the publications are coordinated through the headquarters office.

REIMBURSEMENT POLICY

The three societies will supply stationery and envelopes to journal editors and technical editors. They will, if requested, also reimburse certain editors for postage and secretarial expenses.

The amount allowed for secretarial reimbursement is evaluated annually. As of the 1987–1988 fiscal year, those entitled to ask for secretarial reimbursement were the editor and technical editors of *Agronomy Journal*, the technical editors of *Crop Science*, the editor-in-chief and technical editors of *Soil Science Society of America Journal*, and the editors of *Journal of Environmental Quality, Journal of Agronomic Education*, and the *Journal of Production Agriculture.*

Those requesting secretarial or postage reimbursement should submit their claims quarterly (31 March, 30 June, 30 September, and 31 December).

EDITORS-IN-CHIEF

The ASA Editor-in-Chief is responsible for editorial policy coordination of all ASA publications and serves as chair of the Editorial Policy Coordination Committee of the three societies. This committee consists of the editors-in-chief of the three societies and is concerned with implementing appropriate policies that cover publications of all three societies. The ASA Editor-in-Chief also serves as the chair of the Editorial Affairs, Policies, and Practices Committee of ASA. This committee is concerned with policy coordination within ASA. This committee is composed of the editors of *Agronomy Journal*, *Journal of Production Agriculture*, *Journal of Agronomic Education*, *Journal of Environmental Quality*, and chair of the Monographs Committee.

The ASA Editor-in-Chief is responsible for advising the president, executive committee, and board of directors on all ASA publications including those that are cosponsored by other societies. The ASA Editor-in-Chief is a nonvoting member of the ASA Board of Directors. At meetings of the board of directors, the ASA Editor-in-Chief recommends any editorial policy changes that need board action and advises the board on editorial and publication actions of the three societies.

The ASA Editor-in-Chief serves as a member of all ASA editorial boards and publication committees and advises the president, executive committee, and board of directors on publication proposals of these committees. He/she may be called upon to handle special problems through an appeals process and perform other editorial duties requested by the executive committee or the board of directors.

The editors-in-chief make recommendations to their respective presidents-elect concerning the appointment or reappointment of editors of journals when their terms are about to expire or when vacancies occur for other reasons. The presidents may call upon the editors-in-chief for their advice and recommendations on all publications.

Chapter 2. Journal Management and Procedures

This section is an overview of the present responsibilities and practices in the review, handling, and production of publications of the three societies. Comments applicable to all journals are given first, followed by information on specific journals. Additional details and supplementary information are available through the headquarters office.

PUBLISHING PRIVILEGES

Membership Requirement

For an unsolicited manuscript to be reviewed for or published in *Agronomy Journal*, *Crop Science*, or the *Soil Science Society of America Journal*, one of its authors must be an active member of one of the three societies. (The word "active," as used here, means either active, sustaining member representative, emeritus, graduate student, or dues-paying undergraduate student membership.) As explained later in this manual, the membership requirement for these journals may be waived under certain circumstances for invited papers.

When each unsolicited manuscript is received by one of these three journals, the editor or managing editor will determine if one of its authors is an active member. If so, the review process can begin as outlined for the specific journal. If not, the manuscript is returned to the author with a membership application and a note explaining the membership requirement.

If the review process extends into another calendar year, it is the responsibility of the managing editor, upon receipt of a manuscript approved for publication, to verify that at least one of the authors is an active member for the current year. If not, the manuscript is to be held and the corresponding author informed of the requirement for the member to renew the membership before the manuscript can be published. If the renewal is not received within 60 days, the manuscript is released to the authors. Active membership is not a publishing requirement for *Journal of Agronomic Education*, *Journal of Environmental Quality*, or *Journal of Production Agriculture*.

RECORDS

Headquarters or the journal's editor or editor-in-chief retain the review record and comments for released manuscripts for 1 year after the release date. The editor, technical editors, and associate or consulting editors of the journal should retain records of released manuscripts for 2 years. Headquarters will retain records of approved manuscripts for 13 weeks after their publication. Editors, technical editors, and associate or consulting editors may dispose of their records on published manuscripts 13 weeks after publication.

ABSTRACT PREPARATION

Abstracts are a unique form of literature that require a great deal of care on the part of both the author and the editor. Abstracts should be completely self-explanatory and intelligible in and by themselves, while at the same time be a suitable literary adjunct to the printed article.

The Checklist and Sample Abstract on page 5 was prepared to explain what should and should not be included in an abstract and to demonstrate how this material should be presented.

Quantities of this form are available at no charge to technical and associate editors who want to make use of them in their correspondence with authors who are having difficulty preparing an adequate abstract.

THE REVIEW-EDITING PROCESS

The process of converting a raw manuscript into a published technical paper is complex, involving numerous persons with various areas of expertise. The reason for the process was summarized by O'Connor (1987) with the statement, "Protect readers from authors and authors from themselves."

The first part—protecting readers from authors—involves maintaining the high standards of the publication itself. The first responsibility of technical and associate editors and reviewers involves determining that the manuscript contains material that is suitable for publication where it was submitted. The scientific editing and review process involves making certain that the research project sought information that was previously not known or not completely understood, that the research was properly set up, accurately conducted and recorded, and the results correctly interpreted.

Although the prime responsibility for accuracy and completeness of the paper rests with the author(s), technical and associate editors and reviewers help assure that all the scientific and technical details are complete and correct and that all errors and ambiguities are caught and corrected. In this sense, the review process protects authors from themselves.

Errors and ambiguities can be grouped into two general categories, (i) scientific and technical, and (ii) grammatical. Although there is substantial

Checklist and Sample Abstract

Abstracts accompany all articles in Society journals. They are often republished as printed in secondary abstracting services and journals, such as *Chemical Abstracts* and BIOSIS or *Biological Abstracts*. The abstract, therefore, should meet two requirements. An agronomist reading an abstract should be able to tell quickly the value of the report and whether or not to read it thoroughly. It also should provide the literature searcher enough information to assess its value and to index it for later retrieval. Use the checklist below to assure that your abstract meets these requirements.

The abstract should:

______ Strive for an impersonal, noncritical, and informative account.

______ Give a clear, grammatically accurate, exact, and stylistically uniform treatment of the subject.

______ Provide rationale or justification for the study. The statement should give a brief account of the purpose, need, and significance of the investigation (hypothesis or how the present work differs from previous work).

______ State the objectives clearly as to what is to be obtained.

______ Give a brief account of the methods, emphasizing departures from the customary. Be specific.

______ Name the soil type by the new classification system if soil type is a factor in interpreting the results.

______ Clarify whether it is a greenhouse or field experiment.

______ Identify scientific name of plants, other organisms, and chemicals.

______ State results succinctly.

______ Outline conclusions or recommendations, if any. An emphasis of the significance of the work, conclusions, and recommendations. This may include new theories, interpretations, evaluations, or applications.

______ Use specific figures whenever possible to avoid use of general terms, especially in presenting the method and reporting the results. For example, if two rates of a treatment are used, state what they are.

______ Never cite references.

______ Contain about 200 to 250 words for articles or 100 to 150 words for Notes.

With permission of the author, a published abstract is reproduced here broken apart with its sections labeled. Study it to see both overall construction of the abstract and the contents of each of its parts. [From *Agron. J.* 78:720–726 (1986)].

Dryland Grain Sorghum Water Use, Light Interception, and Growth Responses to Planting Geometry

J. L. Steiner*

ABSTRACT

Rationale — **Crop yields are primarily water-limited under dryland production systems in semiarid regions.**

Objectives — **This study was conducted to determine whether the growing season water balance could be manipulated through planting geometry.**

Methods — **The effects of row spacing, row direction, and plant population on the water use, light interception, and growth or grain sorghum [*Sorghum bicolor* (L.) Moench] were investigated at Bushland, TX, on a Pullman clay loam (fine, mixed, thermic Torrertic Paleustoll).**

Results — **In 1983, which had a dry growing season, narrow-row spacing and higher population increased seasonal evapotranspiration (ET) by 7 and 9%, respectively, and shifted the partitioning of ET to the vegetative period. Medium population crops yielded 6.2 and 2.3 Mg/ha of dry matter and grain, respectively. High population resulted in high dry matter (6.1 Mg/ha) and low grain yield (1.6 Mg/ha), whereas low population resulted in low dry matter (5.4 Mg/ha) and high grain yield (2.3 Mg/ha). Row direction did not affect water use or yield. In 1984, dry matter production for a given amount of ET and light interception was higher in the narrow-row crops. Evapotranspiration was less for a given amount of light interception in the narrow-row crops and in the north-south row crops.**

Conclusions — **Narrow-row planting geometry appears to increase the partitioning of ET to the transpiration component and may improve the efficiency of dryland cropping systems.**

overlap in duties, problems in Category i are the basic responsibility of technical and associate editors and reviewers and those in Category ii are the basic responsibility of the professional editing staff at the ASA Headquarters Office.

Editors, technical and associate editors, and reviewers cannot ignore grammatical problems, however. Poor writing is often an indication of poor thinking or poor scientific research. If a submitted manuscript is nearly illegible or unintelligible, the editor or reviewer should not hesitate to send the manuscript back to the author for improvements before serious scientific review is begun.

Once a manuscript is readable, scientific editors and reviewers give it a thorough review. Specific aspects of this are outlined in the checklist at the end of this section. In addition, there are some items that will be dealt with in more detail.

One is that editors and reviewers make certain the paper conforms to all aspects of style established by the three societies. For example, if SI units are not used where required, editors or reviewers should make, or ask the author(s) to make, the proper conversions. Correct nomenclature is important also; the use of million metric tonnes is not acceptable, for instance; teragrams (Tg) should be used.

Such things as fertilizer rates should be checked; elemental, rather than oxide units, are required for plant nutrients. (Both may be included, with the oxide form in parentheses after the elemental listing.) Plants must have the complete botanical name after their first mention, soils need full identification, animals need scientific names, chemical compounds need descriptions, manufacturers of brand-name products should be identified by full company name, city, and state.

If any of the scientific and technical items are missing, misspelled, or incorrect in any way, calling the error to the author's attention is esesential.

When a percentage is used to denote plant mass or makeup of a solution, the authors should specify if it is on a volume (v/v) or weight (w/w) basis. If this has not been done, the reviewer should see that this error is corrected.

MARKING EDITORIAL CHANGES

Once the paper is scientifically correct and complete, the technical and associate editors should not consider their work completed. They can give invaluable assistance toward final preparation of the manuscript for publication. If retyping is required (as is often the case) or simply if a final clean version will be produced on a word processor, the author should be made aware of all mechanical requirements; particularly the need that all parts of the manuscript—including footnotes, tables and their captions, and the references section—be doublespaced.

Staff editors at ASA Headquarters appreciate assistance with the more detailed aspects of the manuscript. For example, it is not always easy to distinguish an l (lower case letter l) from a 1 (numeral 1) in text or formulas. Technical or associate editors should highlight these with a penciled-in "el" or "one." The 0 (zero) and O (oh) are a similar case. Where an "x" is used as a multiplication sign, it should be so noted.

Mathmatical equations and Greek letters can often present problems, particularly in regard to such things as exact character placement (baseline, superiors, inferiors), identification of variables, constants, and unknown quantities. Single letters that denote mathematical constants, variables, and unknown quantities in text and equations should be in italic type; vectors should be in boldface roman type. Greek letters should be identified at first occurrence by name and by whether they are upper or lower case. Any help the technical or associate editor can give in this regard is not only appreciated at headquarters, it can also eliminate costly typesetting corrections later.

In most cases, a letter or series of letters used to indicate a gene should be italicized. If the author did not underline it (them), headquarters staff editors appreciate it when the scientific editors or reviewers do.

Single quotation marks are to be used the first reference to denote cultivars (unless the word "cultivar" or the intials "cv" have accomplished that task). Single quotes are *not* to be used with cytoplasms, germplasms, or parental lines, however. If the author has made this error, the scientific reviewers should correct it.

Note: Editors and reviewers who make any of the above-recommended additions, changes, notes, or comments on manuscripts are asked to use a medium that can easily be erased. A soft black pencil is preferred. Please do not use pens, markers, or colored pencils.

This request is made because notes and comments are understandable and appreciated by staff editors, but often need to be added to, rewritten, partially replaced, or moved before work on the manuscript is completed. Such changes are very difficult when these notes and comments cannot be erased.

Technical and associate editors and reviewers are not expected to know the dozens of computer-readable codes used to give typesetting instructions. Instead, they use terms familiar to them, and staff editors make the proper translations. For example, reviewers and editors use the word "multi" or "multiplication sign" to indicate that an "x" in the text represents a multiplication sign. This must be replaced with the specific code to tell the typesetter to use the sign. Another example, the symbol "#" is often used by reviewers and editors to indicate the need to insert a space within the text. This symbol has an entirely different meaning to the typesetter, though, and must be removed and replaced with the correct code to indicate insertion of the needed space.

Checklist for Detailed Comments

Title. Does the title adequately describe the subject of the manuscript? Does it have 12 or fewer words? Can the wording be improved?

Abstract. Abstracts are the most widely read section, often being seen without the paper itself. Does the abstract tell in brief what was done and what was found? More information about abstracts can be found elsewhere in this handbook and in the tri-society *Handbook and Style Manual* (1988).

Review of Literature. Is due credit given to relevant contributions? Is the author's contribution placed in its proper perspective in relation to the state of knowledge? Is the number of references adequate, too small, or excessive?

Objectives. Is the statement of objectives adequate and appropriate?

Methods. Are the methods appropriate? Have suitable measurements been performed? Have proper control measurements been made? Have the methods been presented in sufficient detail to allow a competent scientist to repeat the work? If not, are the sources cited where sufficient detail is available?

Clarity. Does the author present the information in a relatively simple, straightforward manner that can be understood by a reasonably competent scientist? Do the author's words say what they are supposed to mean?

Organization. Does the manuscript develop the subject logically and effectively?

Consistence. Are the same spellings, abbreviations, etc. used throughout? If an abbreviation is defined early in the paper, is it used after that? If not, it can be eliminated.

Editorial Style. Does the manuscript conform to current tri-society editorial style and format, including the use of appropriate SI units?

Duplication. Does the manuscript unnecessarily repeat already published work? Can the manuscript be shortened without loss of content by condensing two or more tables into one? Are all figures needed if the same data are also given in tabular form? Is there unnecessary duplication in the text?

Calculations. Reviewers should randomly select a few instances and verify the calculations made by the author.

Effectiveness of Presentation of Data. Would data presented in tables be better presented in figures, or vice versa?

Tables and Figures. Are tables and figures understandable and complete apart from the text?

Sequence of Tables, Figures, and Equations. Are all serially numbered items presented in the proper order?

Titles of Tables and Figures. Do the titles accurately state the content, or could they be improved?

Table Headings. Is the interpretation clear, unequivocal, and in SI units?

Correspondence of Text with Tables and Figures. Are all tables and figures referred to in the text? Do statements in the text correspond to the content of tables and figures?

Graphs. Do they conform to the new guidelines as outlined in the *Handbook and Style Manual* of the three societies (1988)? Do they contain all observations? Is the plotting of the data accurate?

Conclusions. Are they adequate and supported by the data?

Conjecture. Does the author clearly distinguish between fact and conjecture? Is the amount of conjecture excessive?

References. Is the required information there, complete, and in the proper format? Are there obvious errors, such as misspelled names of authors or publications?

Correspondence of Text and Reference. Do all the references have a corresponding citation in the text and vice versa? Is the year of publication in the citation the same as the year given in the reference list? Are the spellings of author name(s), etc. the same in both places?

AMERICAN SOCIETY OF AGRONOMY

Agronomy Journal

Agronomy Journal first appeared in 1910, just 3 years after the ASA was founded. The first four volumes were titled *Proceedings of the American Society of Agronomy*. From 1913 through 1948 the name was *Journal of the American Society of Agronomy*. In 1949 it was changed to *Agronomy Journal*. Published first in annual bound volumes and later at greater frequency, it appeared from 1923 through 1960 as a monthly periodical. It is now published bimonthly.

The editorial board consists of the ASA Editor-in-Chief; an editor; a managing editor; technical editors in crops, soils, agroclimatology, and agronomic modeling; a number of associate editors covering numerous subject-matter areas of the agronomic, crop, and soil sciences; the executive vice president; and the vice president—management and operations.

Articles reporting original research findings in soil-plant relationships; crop science; soil science; biometry; crop, soil, pasture, and range management; crop forage and pasture production and utilization; turfgrass; agroclimatology and agronomic modeling; and computer software may be published in *Agronomy Journal* subsequent to critical review and approval by the editorial board.

Articles must make a significant contribution to the advancement of knowledge or toward a better understanding of existing agronomic concepts. The study reported must be of potential interest to a significant number of scientists. Papers covering experimental techniques, apparatus, and observations of unique phenomena are published as Notes. These are normally less than two printed pages in length. Additionally, the ASA Board of Directors has approved publishing computer programs in *Agronomy Journal*.

Review papers, either by invitation of the editorial board or volunteered, may be printed in the journal. Invitational papers from nonmembers may be published upon approval by the president if they are accepted by the editorial board.

A contribution to *Agronomy Journal* should be prepared according to instructions given in the *Publications Handbook and Style Manual* (1988) and four copies of the manuscript sent to the editor. After membership verification, the paper is sent by the editor to the appropriate technical editor who obtains two or more reviews through associate editors who also participate in reviewing papers. Reviewers remain anonymous to the authors and each other. The technical editor sends papers approved for publication to the managing editor. The managing editor sees to it that the authors are notified of approval of their manuscripts.

Responsibilities

Editor

The editor is responsible for overall quality of the intellectual content of the journal. As chair of the editorial board, the editor implements policy decisions approved by the board of directors on editorial matters. The editor and editorial board oversee procedures for manuscript submission, review and referee criteria, acceptance, release, and publication. The editor delegates editorial functions to other members of the editorial board, and advises on administrative functions of the managing editor. The editor advises the ASA Editor-in-Chief and the ASA President on the composition of the board. The editor takes an active interest in defining the journal's aims, policies, and editorial coverage, and in recruiting reviewers and members for the editorial board. The editor examines the scope of the journal regularly and, with advice of the editorial board, can recommend that the scope be changed. The editor, by broad knowledge of the field, recommends the mix of specialties to be represented on the editorial board. The editor operates an appeal procedure for authors whose manuscripts are rejected. The editor may write editorials or solicit manuscripts on special topics. Letters to the editor are approved by the editor, but the editor seeks advice of the editorial board and others as needed.

The editor prepares an annual report for the society describing the journal status and recommendations for changes, and the minutes of the annual board meeting published each year.

Technical Editor

Under the direction of the editor, technical editors are responsible for the technical and intellectual content of the journal in their assigned areas. They direct the work of assigned associate editors in reviewing and evaluating the worthiness of manuscripts submitted to the journal. They may assign manuscripts to the associate editor for statistics as appropriate or seek guidance from this associate editor on proper use of statistics. Technical editors delegate to associate editors the responsibility of obtaining additional reviews from qualified peer scientists. Technical editors, however, maintain the overall responsibility for determining that in-depth and timely reviews are obtained from knowledgeable scientists. When revisions of manuscripts are needed, technical editors may delegate to associate editors the responsibility of corresponding and working with authors. Technical editors notify authors when manuscripts are found to be unworthy of publication and inform the managing editor and editor of this action. Normally, at least two negative reviews are required before manuscripts are recommended for release and two favorable reviews are required before manuscripts are recommended for acceptance. Technical editors also are responsible for taking an active interest in assisting the editor to define the journal's aims, policies, editorial coverage, and in recruiting members for the editorial board.

Associate Editor

Under the direction of a technical editor, associate editors are responsible for evaluating *in a timely manner* the technical and intellectual content and suitability of manuscripts assigned to them. Associate editors seek the services of qualified peer reviewers. Technical editors may delegate to associate editors the responsibility of corresponding and working with authors to obtain revisions as needed.

Associate editors make a recommendation to the technical editor about a course of action regarding the disposition of assigned manuscripts, but do not approve or disapprove manuscripts and should not convey to the author the impression that the manuscript will be approved or released. When manuscripts are recommended for release, sufficient reason should be given so that the author can be fully informed by the technical editor. Associate editors are also responsible in assisting the editor in defining the journal's aims, policies, editorial coverage, and in recruiting members for the editorial board.

Managing Editor

The managing editor of *Agronomy Journal* is assigned by the vice president—management and operations. The managing editor supervises copy editing of papers approved for publication, typesetting, transmittal of proofs to authors, makeup of pages for printing, and reprints of articles. The managing editor, together with the vice president—management and operations, makes contract arrangements for production of the journal. The managing editor works closely with the editor to maintain the quality of the journal.

Manuscript Handling

General Procedures

Manuscripts are submitted to the editor, who verifies author membership, determines that there are four copies on line-numbered paper, and that the manuscript fits the format requirements of the journal. Each manuscript is given a registration number and a Register and Record of Manuscript form is prepared. The editor notifies the author of receipt of the manuscript, and its registration number, and sends the author a permission to print and reprint form. The date of receipt determines the priority of publication for the approved manuscript. The editor assigns the manuscript to a technical editor on the basis of the general subject area. One copy of the record sheet is retained by the editor and three copies are forwarded to the technical editor.

If a manuscript is outside the technical editor's competence or cannot be judged impartially, it may be transferred to a different technical editor. If it is transferred, the editor is notified.

The technical editor assigns the manuscript to an associate editor based on subject matter speciality. The associate editor receives three copies of the manuscript, a Register and Record of Manuscript form, three Instructions to Reviewers forms, a Reviewer's Report form, and a Checklist and Sample Abstract form. A manuscript should not be sent to an associate editor at the same institution or research group as the author(s) of the manuscript.

The associate editor, upon receipt of the manuscript, reads it and proceeds in one of the following ways:

1. If the manuscript is too far outside the associate editor's area of competence or it cannot be judged impartially, it is returned to the technical editor for reassignment.
2. If the manuscript is so poorly prepared that it would obviously be rejected, the associate editor returns it directly to the technical editor with the appropriate comments.
3. A manuscript that does not fall into categories (1) or (2), is sent to one or two competent peer reviewers along with a copy of the preprinted instructions for reviewers. The reviewer's name and the date sent are recorded on the record sheet. The associate editor should always act as one of the reviewers.

When the associate editor receives the reviewers' comments, the date received is recorded on the manuscript record form and the manuscript and the reviewers' comments evaluated. The associate editor then takes one of the following actions:

1. If the reviewers recommend publication without change and the associate editor's review agrees, the associate editor submits the manuscript and all pertinent materials to the technical editor. Such instances are rare.
2. If the reviewers and the associate editor believe that a manuscript could be published after some revisions and the technical editor in-

volved has delegated responsibility to do so, the associate editor corresponds with the author to obtain a satisfactory revision. If the technical editor has not delegated responsibility for corresponding with the author, the associate editor forwards the reviews to the technical editor who corresponds with the author. In the correspondence with the author, the associate/technical editor should acquaint the author with the rule that manuscripts held for revision longer than 26 weeks are subject to release (see below). The associate/technical editor should not give the author the impression that the manuscript will be accepted for publication if the manuscript is revised. If the associate editor corresponds with the author and receives a revised manuscript, two copies of the revised manuscript, one copy of the original manuscript, one copy of all reviewers' comments, one copy of all correspondence, one copy of the record sheet, and a recommendation are forwarded to the technical editor. Associate and technical editors keep file copies of all correspondence. The section in this handbook on records provides guidelines on how long files should be kept.

3. If the technical editor approves the manuscript for publication, the author is notified and one copy of the revised manuscript, one copy of reviewers' comments and correspondence, and the completed Register and Record form are forwarded to the managing editor. The author will be notified of probable publication date in the query letter that accompanies the proofs.
4. If reviewer(s) and the associate editor recommend release of a manuscript, the associate editor forwards to the technical editor copies of all correspondence concerning the manuscript, copies of the manuscript with reviewers' comments written on them, one copy of the record sheet, and two copies of a brief statement outlining the basic reasons for releasing the manuscript. The associate editor does not correspond with the author. When a manuscript is released, the technical editor notifies the author and a blind copy of the letter of release is sent to the appropriate associate editor. The technical editor returns a copy of the manuscript and copies of the reviewers' reports to the author. The editor receives a copy of the letter of release and a copy of the reviewers' reports. Releases may be appealed to the editor.
5. If one reviewer recommends release and another recommends acceptance, the associate editor proceeds as outlined above on the basis of the associate editor's evaluation. Such situations are sensitive and require careful handling.

Technical editors can approve or release manuscripts against the majority opinion of the reviewers and the associate editor.

If manuscripts sent by the associate/technical editor to an author for revision are not returned within 26 weeks, they are released by the technical editor unless extenuating circumstances exist. Once released, manuscripts must be resubmitted to the editor to receive additional consideration by the journal.

Preparation of Papers for Publication

Copyediting

Approved manuscripts are prepared for printing according to their dates of original submittal and final approval and according to the space available in the next issue of *Agronomy Journal.*

Each manuscript is read by the managing editor or an assistant editor to correct typographical, grammatical, and style errors and to improve the writing where possible and appropriate, with special care not to change the meaning. Instructions for the typesetter are inserted as needed.

Tables are set in type in the proper form for reproduction, and drawings and photographs for the figures in the paper are scaled according to their dimensions, the size of lettering, and other factors.

Transmittal of Proofs to Authors

The proofs received from the typesetter are measured to determine number of pages the article will require and are then sent to the authors along with the manuscript, figure proofs, table proofs, and a reprint order form showing estimated page and publication charges. Authors are responsible for proofreading this material.

Return of Proofs to Typesetter and Preparation of Page Dummy

Corrections and changes indicated by the authors on their proofs are transferred to the master proofs, which are then returned to the typesetter.

A paste-up dummy, showing the layout of the pages of the journal, is made and sent to the typesetter. The running heads are indicated on the dummy and a table of contents is prepared and sent with it.

Correction of Page Proofs, Camera-Ready Pages, and Printing

The page proofs supplied by the typesetter are compared with the dummy pages and the galley proof corrections. Necessary corrections are marked on the page proofs. Final camera-ready pages are prepared for the printer, who makes negatives of them for manufacture of the printing plates. A positive copy (silver or blue print) of these negatives is inspected at headquarters, then returned to the printer along with an order for the number of copies to be printed.

Preparation of Reprints

The production and shipping of reprints is managed by the headquarters office based on information from the Reprint Order and Publication Charge forms completed by each author.

Journal of Agronomic Education

The *Journal of Agronomic Education*, established by ASA in 1971, is published twice a year. It is an outgrowth of the Agronomic Education section formerly carried in *Agronomy Journal*. The journal accepts reports of original studies pertaining to concepts of resident, extension, and industrial education in plant and soil sciences. Other types of manuscripts published in JAE include letters to the editor, slide set articles, computer software papers, ideas, news features, and book reviews. Analysis and synthesis of existing knowledge or research, instructional techniques and methodology, surveys of instruction, and other studies that contribute to the development or better understanding of educational efforts are encouraged. Reviews or digests of a comprehensive and well-defined scope are acceptable. Articles may confirm and strengthen the findings of others, revise established ideas or practices, or challenge accepted theory, providing the evidence presented is significant and convincing. Manuscripts based chiefly on personal philosophy or opinion are acceptable if they conform to the above criteria.

Authors of papers submitted to *Journal of Agronomic Education* are not required to be society members.

The editorial board consists of the ASA Editor-in-Chief, an editor, a number of associate editors, a managing editor, the executive vice president, and the vice president—management and operations.

Responsibilities

Editor

The editor is responsible for the overall quality and intellectual content of the journal. The editor supervises the activities of the associate editors and makes final decisions on acceptance or release of manuscripts. If a contribution is released, the editor corresponds directly with the author, stating reasons for the rejection.

The editor serves as chair of the editorial board of the journal, and works with members of the editorial board to recommend editorial policies to the board of directors. The editor informs the board of directors of matters through the editor-in-chief and maintains close contact with the managing editor and the headquarters office on production of the journal. Other responsibilities are similar to those shown for the editor of *Agronomy Journal*.

Associate Editors

Selected in fields of specialization according to the scope of *Journal of Agronomic Education*, associate editors are appointed in resident education for soils and crops, extension education for soils and crops, industrial education, international education, and general extension education. Associate editors solicit reviews, evaluate papers, and recommend acceptance, revision, or release. Their responsibilities and those of reviewers are similar to those listed for *Agronomy Journal*. If an associate editor believes that a manuscript should be released, the entire file should, without contacting the author, be returned to the editor for further processing.

Managing Editor

Duties of the managing editor are similar to those outlined for *Agronomy Journal.*

Manuscript Handling

A paper intended for *Journal of Agronomic Education* must be prepared according to the *Publication Handbook and Style Manual* (1988) and sent in quadruplicate to the managing editor. The journal regularly publishes a Suggestions to Contributors, which provides requirements for manuscript preparation and submission. The managing editor maintains a log of all manuscripts submitted for publication in JAE. A copy of the log is sent to the editor occasionally for record-keeping purposes. The managing editor initiates a Register and Record of Manuscript form, assigns a number to the manuscript, and sends three copies of the paper to an associate editor selected by the editor, along with a Register and Record of Manuscript form, and three copies of the Instructions to Reviewers for Manuscripts, Review Procedure form, Suggestions for Contributors forms, and Reviewers Report form. The associate editor will arrange for at least two reviews. At the time the paper is forwarded to the associate editor, the managing editor also forwards a Permission to Print and Reprint form to the author. At the time of publication for each issue, the managing editor sends the editor a status report of the manuscripts published in the upcoming issue and the manuscripts in review. The remaining steps toward publication are similar to those for *Agronomy Journal.* Some variations may be used because of the differences in frequency, size, and method of publication of the two journals. Appeals of editorial decisions are handled by the ASA Editor-in-Chief.

MULTI-SOCIETY JOURNALS

Journal of Environmental Quality

The *Journal of Environmental Quality* is published jointly by the American Society of Agronomy, Crop Science Society of America, and the Soil Science Society of America on a quarterly basis . The first issue was published in January 1972. The editorial board consists of the editors-in-chief of ASA, CSSA, and SSSA; an editor; a number of associate editors and consulting editors to handle subject matter areas: a managing editor, the executive vice president; and the vice president—management and operations.

Authors of papers submitted to *Journal of Environmental Quality* are not required to be members of the three societies. Contributions reporting original research or brief reviews and analyses dealing with some aspect of environmental quality in natural and agricultural ecosystems are accepted from all disciplines for consideration by the editorial board. Papers may be volunteered, invited, or coordinated as a symposium. Acceptance of a group of symposium papers for collective publication is limited to one such group per issue. Book reviews may be invited by the editor. The two main sections in the journal are Technical Reports, containing original research papers, and Reviews and Analyses, containing appropriate papers.

Papers in *Journal of Environmental Quality* cover various aspects of anthropogenic impacts on the environment, with particular focus on terrestrial and aquatic systems. Emphasis is given to the understanding of underlying processes rather than to monitoring, and papers should be broad in scope. Topics receiving attention in the Journal include: reutilization of agricultural, municipal and industrial wastes; transport, transformations, fate, and ecological impact of nutrients, pesticides, and toxic organics; impacts of acid precipitation on plants, soils, and aquatic biota; and reclamation and management of drastically disturbed lands.

A paper intended for *Journal of Environmental Quality* must be prepared according to the *Publications Handbook and Style Manual* (1988) and the "Suggestions to Contributors" frequently published in the journal. Four copies of each manuscript are sent to the editor. The editor sends the paper to an associate editor or a consulting editor, who will arrange for at least two reviews by peer scientists. Authors have 26 weeks in which to revise manuscripts. The remaining steps toward publication are similar to those for *Agronomy Journal*. Appeals of editorial decisions are handled by the editor-in-chief of ASA.

Responsibilities

Editor

The editor serves as chair of the editorial board for the journal and supervises the editorial activities of the associate and consulting editors. The editor interacts with members of the editorial board to set procedures for manuscript submission, review, referee criteria, acceptance, release, and publication. The editor solicits suggested topics and authors for invited review or synthesis papers and extends the invitation to prospective authors. As requested, the editor reports to the boards of directors of the three societies on the state of the journal and on the activities and recommendations of the editorial board; the editor also keeps the ASA Editor-in-Chief informed about editorial progress and maintains close contact with the managing editor and the headquarters office with regard to the production of the journal. Other responsibilities are similar to those shown for the editor of *Agronomy Journal*.

Associate and Consulting Editors

The associate editors and consulting editors are chosen according to subject matter areas of environmental quality instead of the traditional subject matter divisions of the three societies. Associate editors are members of one of the three the societies, consulting editors are not members; both have the same duties and responsibilities. Their responsibilities, and those of the reviewers are similar to those listed for *Agronomy Journal*.

Managing Editor

Duties of the managing editor are similar to those outlined for *Agronomy Journal*.

Manuscript Handling

The editor receives four copies of all manuscripts and examines the titles and contents to detect any obvious misdirection of a paper or serious omission of material. All papers suitable for review are registered as follows: The editor (i) records date of receipt; (ii) assigns a registration number to the paper which is determined serially by the date of receipt; and (iii) specifies whether the manuscript is a Technical Report or Review and Analyses paper. A copy of the registration information may be sent to the ASA Editor-in-Chief for review as desired. The corresponding author is notified of the receipt of the paper and the registration number, which must be used in all correspondence regarding the manuscript. The editor assigns each manuscript to an associate or consulting editor according to the field of specialization.

The procedures are essentially the same as described for *Agronomy Journal*. Some variations may be used because of the differences in frequency, size, and method of publication of the two journals. One difference is that the Permission to Print and Reprint form is sent to the authors with the proofs instead of earlier.

Journal of Production Agriculture

The *Journal of Production Agriculture* began quarterly publication in 1987. It is published jointly by ASA, CSSA, and SSSA. The American Forage and Grassland Council, Weed Science Society of America, American Society of Animal Science, American Agricultural Economics Association, and Society for Range Management are cooperators in the journal. The editorial board consists of the editors-in-chief of ASA, CSSA, and SSSA; an editor; a number of associate and consulting editors to handle the subject matter areas; a managing editor; the executive vice president; and the vice president—management and operations.

The editorial board is composed of three associate editors from each of three areas of general agronomy, soils, and crops, and one consulting editor from each of the cooperating societies. Additional members may be added at a later date if needed.

Authors of papers submitted to *Journal of Production Agriculture* are not required to be members of the three societies. Contributions to the journal may be papers of original research, interpretive review or state of the art papers, notes on original research or new research techniques or equipment, and comments and letters to the editor. The journal's aim is to transfer production-oriented information in a manner that will be of use to a wide range of professional agriculturalists. All manuscripts must be in the Customary (inch-pound) system of measurement unless a commonly accepted unit does not exist for the character being measured. Inclusion of the chemical formulation along with the common or generic name of chemicals used is encouraged. Each feedstuff named in the manuscript may be identified by the International Feed Number (IFN) and a simplified name of the feed.

Emphasis is placed on publishing papers that transfer production-oriented scientific and technological information on agronomy in concert with other disciplines, such as animal science, weed science, agricultural economics, entomology, plant pathology, horticulture, and forestry. Articles should contain an interpretive summary. This may be included either within the contents of the discussion section or as a separate summary at the end of the manuscript.

Synthesis or interpretive review articles must provide a comprehensive review of appropriate papers on a given subject. Emphasis should be placed on interpreting the information for its applicability, both positive and negative, for a wide range of conditions. Review articles should be oriented toward an applied problem or opportunity. Review articles may be invited by the editor after concurrence with the editorial board or they may be submitted on a voluntary basis. Research articles must be reports of original research emphasizing applied and interdisciplinary studies. Studies that do not lend themselves to the conventional scientific approach, but are of considerable significance in practical agriculture, are welcomed. All manuscripts must report the outcome of scholarly inquiry and should stress practical applications.

Except for the exceptions already stated, papers intended for *Journal of Production Agriculture* must be prepared according to the *Publications Handbook and Style Manual* (1988) and four copies sent to the editor. The editor sends the paper to an associate or consulting editor who arranges for at least two reviews. Authors have 26 weeks in which to revise the manuscript. The remaining steps toward publication are similar to those for *Agronomy Journal*, with some variation due to the size and frequency of the two journals. Appeals of editorial decisions are handled by the ASA Editor-in-Chief.

Responsibilities

Editor

The editor serves as chair of the editorial board and is responsible for the overall quality of the journal and implementing policy made by the ASA Board of Directors, with concurrence of the CSSA and SSSA Boards of Directors. The editor supervises the editorial activities of the associate and consulting editors. The editor interacts with members of the editorial board to set procedure for manuscript submission, review, referee criteria, acceptance, release, and publication. The editor in consultation with the editorial board may write editorials or solicit manuscripts on special topics. Other responsibilities are similar to those of the editor of *Agronomy Journal*.

Associate and Consulting Editors

The associate and consulting editors are chosen according to subject matter areas. Consulting editors are nominated by and represent the cooperating societies. Associate editors are required to be members of the three societies, consulting editors are not. Both have the same duties and responsibilities.

Associate and consulting editors are responsible for obtaining a minimum of two reviews in addition to their own, for each assigned manuscript. To expedite review, these editors are encouraged to contact respective reviewers by telephone before forwarding manuscripts and to request return of reviewed manuscripts within 2 weeks. Unless extensive revision is required, revised manuscripts should not be sent out for additional review. Minor revisions should be handled by telephone rather than by returning the paper to the author. Additional editorial responsibilities are similar to those listed for *Agronomy Journal.*

Managing Editor

Duties of the managing editor are similar to those outlined for *Agronomy Journal*, with some variation due to the size and frequency of the two journals.

Manuscript Handling

Four legible copies of each manuscript are submitted to the editor. The editor will assign a registration number to the manuscript, notify the corresponding author of receipt, and forward a Permission to Print and Reprint form if one was not enclosed. The editor will assign the manuscript to an associate or consulting editor on the basis of subject matter. The remaining procedures are essentially the same as described for *Agronomy Journal.* Some variations may be used because of the differences in frequency, size, and method of publication of the two journals.

CROP SCIENCE SOCIETY OF AMERICA

Editor-in-Chief and Editor

The editor-in-chief, who is appointed by the CSSA President after ratification by the CSSA Board of Directors, serves as chair of the CSSA Committee on Editorial Affairs, Policies, and Practices; as an ex officio member of all CSSA editorial committees; and as a representative of the CSSA board on the three societies' Editorial Policy Coordinating Committee, and is involved in other intersociety publication activities. The editor-in-chief also may serve as editor of *Crop Science* at the discretion of the CSSA Board of Directors.

The editor of *Crop Science* serves as chair of the editorial board and keeps in contact with its members regarding policy and procedures, presides at the annual meeting of the board, and consults with the managing editor on printing and other matters. The editor submits an annual report of activities and recommendations to the CSSA Board of Directors and to the CSSA Executive Committee.

Crop Science

Crop Science, published bimonthly, is the official publication of the Crop Science Society of America. Publication began in January 1961, 5 years after CSSA was organized. The publication is prepared by an editorial board consisting of the editor-in-chief, editor, technical editors, a number of associate editors, a managing editor, the executive vice president, and the vice president—management and operations.

Crop Science publishes original research in crop genetics, breeding, cytology, physiology, management, ecology, quality, utilization, pest management, and cell biology and molecular genetics. Critical reviews also may be published. Cultivar, germplasm, and parental line registrations are published upon recommendation by a crop registration committee. Papers covering experimental techniques, apparatus, and observations of unique phenomena are published as notes. These are normally less than two printed pages in length.

Reviews, research reports, or papers that combine the two, and symposium papers may be submitted for publication. They will be subject to normal review procedures and, if accepted, will be published without special recognition, other than a statement in a footnote.

Manuscripts submitted to *Crop Science* must be prepared according to the *Publications Handbook and Style Manual* (1988). *Crop Science* issues frequently include Instructions to the Authors that have the most recent requirements for manuscript preparation and submission. (For further details, see CSSA Publication Policy, usually published in the January–February issue of *Crop Science* each year.)

The Duties of the editor, technical editors, associate editors, and managing editor are similar to those holding those assignments on *Agronomy Journal*, with the differences outlined in the next section.

Manuscript Handling

Four copies of research paper and note manuscripts must be submitted to the managing editor at the headquarters office. Registration articles should be submitted to the registration committee. The masthead of each issue gives instructions in this regard and current addresses. For research papers and notes the managing editor is responsible for verification of membership, assigning a number to the manuscript, initiating a Register and Record of Manuscript form, assigning the manuscript to a technical editor based upon the general subject area, and notifying the author of receipt of the manuscript and its registration number. A Permission to Print and Reprint form is sent at the same time.

Manuscripts that bridge two or more disciplines, such as physiology and genetics, are handled by one technical editor, but are referred to the most appropriate associate editor in other specialties for review of technical content and general conclusions.

The managing editor forwards all copies of the manuscript and all but one copy of the Register and Record of Manuscript form to the technical editor. The technical editor then assigns the manuscript to an associate editor based on subject matter. The technical editor sends the associate editor four copies of the manuscript; three copies each of the Instructions to the Reviewers and Reviewers Report forms, and the Checklist and Sample Abstract form, and two copies of the Register and Record of Manuscript form. Each month the managing editor sends a summary of manuscripts received to the technical editors and the editor.

The associate editor sends a manuscript and a copy of the Instructions to the Reviewers form to at least two competent peer reviewers who have been contacted in advance by telephone. The associate editor writes the return date on the Instructions to Reviewers form, giving the reviewer 2 weeks to return the manuscript to the associate editor. The associate editor should also be a primary reviewer unless the manuscript is outside the associate editor's area of expertise. If the associate editor cannot act as a primary reviewer, the manuscript should be sent to three outside reviewers at the outset. In the interest of timeliness the decision will generally be made on the basis of the first two reviews received that agree. If the reviewers recommend publication of the manuscript with appropriate revision and the associate editor agrees, the associate editor corresponds with the author to obtain satisfactory revision. To avoid unnecessary delay in the review-revision process, associate editors should not seek outside review of major revisions unless the paper is outside the associate editor's area of expertise, and should handle minor revisions by telephone rather than returning the paper to the author.

Associate editors make a recommendation to the technical editor about a course of action regarding the disposition of assigned manuscripts but do not approve or disapprove manuscripts and should not convey to the authors the impression that the manuscript will be approved or released. When manuscripts are recommended for release, sufficient reason should be given so that the author can be fully informed by the technical editor. Technical editors have authority to release or to approve manuscripts for publication. Rejections may be appealed to the editor.

Authors have 13 weeks to complete revisions. Other editorial responsibilities and procedures are similar to those outlined for *Agronomy Journal* except that copies of the letter of release and the Register and Record of Manuscript form are sent to the managing editor and the editor.

Associate editors are also responsible for assisting the editor in defining the journal's aims, policies, editorial coverage, and in recruiting members for the editorial board.

SOIL SCIENCE SOCIETY OF AMERICA

Editor-in-Chief

The editor-in-chief is chair of the editorial board of *SSSA Journal*, is responsible for all publications of SSSA, is a nonvoting member of the SSSA Board of Directors, and is a member of the tri-society Editorial Policy Coordinating Committee. The editor-in-chief serves the dual role of overseeing all SSSA publications and is responsible specifically for *SSSA Journal*. The editor-in-chief receives from authors all manuscripts submitted to the journal, maintains a log of all actions on all manuscripts, and assigns them to a technical editor who is responsible for papers submitted for review in a given society division. The technical editor sends the manuscripts to associate editors for review. If an associate editor judges that a paper is not suitable for publication or requires drastic revision, it is returned to the technical editor, who determines its disposition and communicates with the corresponding author.

The editor-in-chief also maintains contact with members of the editorial board regarding policy and procedures, prepares an annual report, presides at the annual meeting of the board, reports to the SSSA Board of Directors on the condition of the journal and recommendations of the editorial board, and consults with the managing editor on printing and other matters.

Soil Science Society of America Journal

The *SSSA Journal* is the official publication of the Soil Science Society of America. It was first published in 1937, 1 year after the SSSA was organized, and appeared as an annual bound volume through 1951. In 1952 it became a quarterly publication and since January 1958 it has appeared bimonthly. The first 39 volumes were titled *Soil Science Society of America Proceedings*, and in 1976 the name was changed to *Soil Science Society of America Journal.*

The responsibility for upholding editorial standards and policies approved by the SSSA Board of Directors rests with the editorial board of SSSA. The editorial board consists of an editor-in-chief, a managing editor, a number of technical and associate editors including at least one representative for each division of the society, the executive vice president, and the vice president—management and operations. The editor-in-chief is nominated by the president and confirmed by the board of directors. Technical and associate editors are appointed by the president on recommendation of the editor-in-chief.

SSSA Journal is the normal channel for publication of papers reporting original research findings, reviews of research, and comment and letters to the editor as defined in the SSSA publication policy. Papers of appropriate subject matter usually less than two printed pages in length may be submitted as Notes. Invited papers by nonmembers, with prior approval of the SSSA President, may be published in *SSSA Journal* if accepted by the editorial board.

Contributions to *SSSA Journal* must be prepared according to the instructions given in the *Publications Handbook and Style Manual* (1988) with four copies forwarded to the editor-in-chief. Manuscripts that require drastic revision may be released by the editor-in-chief. If resubmitted, they will be given a new date of receipt. For further details, see SSSA Publication Policy, usually published in the January-February issue of *SSSA Journal* each year.

Technical Editors

Under the direction of the editor-in-chief, technical editors are responsible for directing the review of the manuscripts in their assigned divisions. They direct the work of assigned associate editors in reviewing and evaluating the worthiness of manuscripts submitted to the journal. Technical editors delegate to associate editors the responsibility of obtaining additional reviews from qualified peer scientists, corresponding and working with authors, and approving manuscripts which are deemed worthy of publication. Technical editors correspond with authors whose manuscripts need major revisions and decide when manuscripts are considered unworthy of publication, notifying the authors and the editor-in-chief of this action. Normally, at least two negative reviews are required before manuscripts are released and two favorable reviews are required before manuscripts are recommended for acceptance. Technical editors also are responsible for taking an active interest in assisting the editor-in-chief to define the journal's aims, policies, editorial coverage, and in recruiting members for the editorial board.

Associate Editors

An associate editor receives from the technical editor manuscripts which cover subject matter in the associate editor's division. The associate editor obtains at least two reviews for each paper, one of which can be by the associate editor. If two reviews are favorable, the associate editor has the authority to approve manuscripts and send them to the managing editor for publication. If the paper requires minor revisions, the associate editor communicates with the author to obtain it. If the revision is satisfactory, the associate editor may accept the paper for publication. Copies of all correspondence with the authors including the registration number should be sent to the technical editor. If a paper requires drastic revision or is unacceptable after two unfavorable reviews, it is sent to the technical editor. Responsibilities of *SSSA Journal* reviewers are the same as those of *Agronomy Journal* reviewers.

Manuscript Handling

Registration of Manuscripts

Four copies of all manuscripts should be mailed to the editor-in-chief at the address given in the masthead of each issue. The editor-in-chief ascertains the membership status of the author(s) and records the date received. This date determines the priority of publication for the paper after approval.

The editor-in-chief assigns a registration number to the paper, which is determined serially by the date of receipt, and initiates a Register and Record of Manuscript form. Then all copies of the manuscript and the Register and Record of Manuscript form are sent to the appropriate technical editor depending on the division in which the manuscript will be reviewed.

Review of Manuscripts

The technical editor sends the manuscript to an associate editor who obtains at least two reviews of it. A paper sent to its author(s) for revision and held more than 26 weeks will be released and given a new date of receipt if resubmitted. When the associate editor is satisfied with the manuscript, the manuscript is approved for publication and a copy is sent to the managing editor, including the signed original copy of the Register and Record Form, the name and address of the corresponding author, and the dates the manuscript was received and approved. A copy of the Register and Record form should be sent to the technical editor. Manuscripts not approved are sent to the technical editor for final disposition. Authors of such manuscripts should contact the SSSA Editor-in-Chief if they wish to appeal the decision.

Preparation of Papers for Publication

Upon receiving an approved manuscript, the managing editor determines if the manuscript format conforms to the editorial requirements of the journal. A form letter is then sent to notify the corresponding author of receipt of the approved manuscript, a tentative date of publication, the date proofs will be ready, estimated publication charges, and requesting any additional information or material needed. The balance of the procedures are similar to those for *Agronomy Journal*. The main difference is that the Permission to Print and Reprint form is sent to the authors with the proofs instead of earlier.

REFERENCE

O'Connor, M. 1986. How to copy edit scientific books and journals. ISI Press, Philadelphia, PA.

Chapter 3. Procedures for Monographs, Books, and Other Publications

Generally proposals for publishing monographs, books, and other publications are generated within the Societies themselves. Unsolicited or non-Society generated proposals may be considered with a resulting decision based on the following factors:

1. The subject matter is within the technical scope of the society(ies).
2. The initiative for the publication originated from a member of the society(ies).
3. The society(ies) must have the prerogative of naming the editorial committee or editor.
4. The publication is not acceptable for publication by commercial publishers.
5. There is a reasonable chance of breaking even financially on the publication in the long term.

As a token of appreciation for their efforts, each author and editor of a monograph, book, or special publication receives a free copy of the published version. In addition, 200 reprints of each chapter are given to or divided among the author(s) of that chapter.

MONOGRAPHS

The word "monograph," used by the three societies, means a detailed, scholarly treatise written by experts on a single topic or on a particular subject, usually of narrow scope. Publication of a monograph brings recognition to the societies and to the persons who have contributed chapters. A monograph must be scholarly and suitable for the intended reading audience. Beginning in 1949, the first six monographs were published by Academic Press. The remaining monographs have been published by the American Society of Agronomy or jointly published by the three societies.

Agronomy monographs are published on an irregular basis and are prepared by specialists after careful study. A monograph typically is a compilation of the findings of several scientists in a specific research area. All monographs, whether they are revisions of earlier studies or new findings, are planned and revised by committees of qualified society members.

Monographs Committee

A tri-society monographs committee reviews suggestions for new monographs, and revision of existing monographs, and monitors monographs in progress. One of the members of the monographs committee is designated as chair. The editors-in-chief of the three societies serve as ex-officio members of the committee, as does a representative from the headquarters office.

When a topic for a new monograph or revision of an existing monograph is proposed, the monographs committee discusses the topic either by mail or at the annual meeting of the three societies. If the committee agrees that the subject merits further study, it may evaluate the proposal and submit its recommendation to the ASA President or it may recommend three to six competent specialists in the field to function as a monograph feasibility committee. If the ASA President appoints such a committee, it reports through the chair to the ASA President:

1. The need for the proposed monograph and the amount and quality of research information available;
2. The existence of, or plans for, competing books; and
3. The groups that would be interested in purchasing the book and potential sales volume.

The ASA President refers the report of the feasibility committee to the monograph committee. Based upon this report, the monographs committee may recommend to one or all three executive committees or boards of directors the preparation of a monograph. The executive committee(s) or board(s) make the final decision on publication and establish a budget for those approved.

Upon recommendation of the monographs committee chair, the ASA President appoints an editor and editorial committee to develop the monograph. The monographs committee is kept informed of progress and reviews the brief monograph contents outline from the editor. Otherwise, the preparation, review of detailed chapter outlines, and editing of the monograph is the responsibility of the editor and editorial committee.

Duties of the Editor and Editorial Committee

The editor and editorial committee are responsible for the preparation, review, and editing of the monograph. General duties include the determination of subject matter and scope of coverage for the intended audience, organization of subject matter and elimination of duplication, recruitment of the best qualified authors, and assurance of uniformity in style, setting of deadlines, and technical editing of the manuscripts. The editorial committee should obtain the services of well-qualified, dependable authors who are willing and able to complete manuscripts within the specified time limits.

The editor and editorial committee are recognized by having their names listed on the title page of the monograph.

The specific duties of the editor and editorial committee are as follows:

1. Make tentative outlines of headings and subheadings of chapters for the monograph and submit these to the monographs committee for suggestions and approval.
2. Select an author for the monograph or one or more authors for each chapter. Editors may also be authors, although this is not encouraged or preferable because of the added burden on the editor. Moreover an author may prepare or help prepare more than one chapter. The monographs committee and executive committee may review the list of authors.
3. Provide headquarters with a list of chapter authors, complete with affiliation, mailing address, and office telephone number. Distribute the list to all authors to facilitate communications.
4. Advise authors about the desired scope of coverage and the intended audience for the monograph.
5. Apprise each prospective author of responsibilities relating to the time schedule for completion of manuscripts, including figures and tables, and obtain written confirmation from each author of his/her willingness to prepare a manuscript *by a specified deadline.*
6. Obtain a detailed outline of the chapter from each author.
7. Eliminate duplication and organize chapters to make a well-coordinated monograph, prepare an overall outline for the monograph, and send this information to each author, the monographs committee, and the headquarters office.
8. Provide authors with instructions, developed jointly with the headquarters office staff, including copies of the *Publications Handbook and Style Manual* (1988), if needed, to ensure uniformity in citing references, referencing crops, writing style, preparation of graphs and figures, and other parts of the manuscripts.
9. Set deadlines for approval of outlines, first drafts, and other necessary stages of preparation. Inform authors of deadlines and advise them that authors who miss these deadlines or who submit unacceptable manuscripts will be replaced.
10. Complete any forms required by headquarters to be used in publication planning or production.
11. Review and edit manuscripts and make suggestions for coordination, elimination of duplication, uniformity of style, and incorporation of topics needed to make the monograph complete. The editorial committee implements its own manuscript review process, like that required by a society journal, independent of the institutional peer review and clearance obtained by individual chapter author. Generally the editorial committee would want the assistance of outside peer reviewers in editing the manuscripts for scientific and technical content. However, it is important that at least one person on the editorial

committee read all of the manuscripts to assure uniformity of coverage for the monograph.

12. Submit all manuscripts (as a complete monograph, rather than as individual chapters) to ASA Headquarters for style editing.
13. Submit to the managing editor the final chapter outline and a contributor list with their professional titles, department, affiliation, and location.
14. Meet, if possible, with the managing editor for a final discussion after the style editing has been completed by headquarters office staff.
15. Prepare a preface for the monograph.
16. Prepare a subject index for the monograph either (i) by providing a set of index words and phrases to the authors for use in indexing their chapters and subsequently by consolidating these indexes, or (ii) by taking sole responsibility for the entire index. (See section on index preparation.)
17. Assist the managing editor with any problems met in the final preparation of the monograph.
18. Enforce the editorial policies and regulations found in the *Publications Handbook and Style Manual* (1988) and clear any desired changes in regulations with the vice president—management and operations or editors-in-chief.
19. Promptly submit progress reports on a quarterly basis upon request by the monographs committee.
20. Offer suggestions to the headquarters office for promotion of the printed book.

Responsibilities of Chapter Authors

Authors are responsible for the preparation of complete, up-to-date, and definitive chapters covering the general subject matter assigned by or agreed upon with the editor. The author is responsible for including and presenting all pertinent information in the chapter. While the monograph is not intended to contain a complete bibliography of a particular subject, it is essential that omission of a paper from citation not result in incomplete or inaccurate presentation of the subject.

Authors of chapters are responsible for the interpretation of the published literature. They may accept the original author's interpretation, may creatively extend or amplify the original author's interpretation, or may refute or qualify it if other reports or facts support this action. Since a chapter should be a definitive treatment of the subject, authors should devote special attention to situations where there are contrasting or conflicting reports. If information available to the author permits, the author should not hesitate to make a judgment as to the merits of conflicting reports. Conversely, if existing evidence exemplifies unresolved issues, the author should illustrate them.

Authors must obtain reviews of chapters from well-informed colleagues within or outside the author's institution or place of employment before submitting first drafts of manuscripts to the editor. Authors must also comply with the internal manuscript review policy for their institution or agency and provide confirmation of this compliance to the monograph editor. Authors must inform the editor of the persons who have reviewed the manuscripts so that the editor will not subsequently send it to a previous reviewer during the editorial committee's independent manuscript review procedures.

The author must agree that the material in the manuscript will be published first by the three societies and that the societies, as publishers, will control its subsequent distribution.

Each author is expected to submit the following material to the editor according to an established schedule:

1. Detailed outline of the chapter.
2. First draft of the manuscript in duplicate, typewritten, doublespaced, and with lines numbered on each page. (Line-numbered paper is available at no cost from ASA Headquarters.)
3. Final draft of the manuscript, prepared as in (2), after making any changes required by the editor and peer reviewers chosen by the editorial committee.
4. Written permission from the owners to use any copyrighted material found in the chapter—including quotations, pictures, graphs, or tables. It is the responsibility of authors to obtain permission in writing for use of any copyrighted material used in the chapter. A sample form for use in this connection is given in Chapter 8 of the *Publications Handbook and Style Manual* (1988). This form, complete with appropriate signatures from the owner of the copyrighted material (usually the publisher), must accompany the chapter manuscript when sent to the editor.

The authors' work is not completed with submission of their chapter to the editor; authors are expected to work closely with headquarters during the final stages of preparation. The following list outlines the basic areas of responsibility:

1. Provide prompt answers to all queries concerning their chapter during the editing process.
2. Thoroughly and promptly read the typeset proofs of their chapters and mark all corrections that are needed.
3. If requested by the editor, to receive the corrected proofs and use them to prepare the subject index for their chapter (see section on index preparation). Ideally all errors will have been corrected during the first round of proofreading, so additional proofreading at this stage should not be necessary.

The editor will indicate to each author the procedures to be followed to ensure uniformity in style of writing for the text, literature references, illustrations, etc. Each author, however, is responsible for submitting a manuscript in final form, complete with publishable figures and graphs, just as for a professional journal.

Authors must use the *Publications Handbook and Style Manual* (1988) as the official guide for preparing their manuscripts. When the requirements differ for the three societies, those given for ASA publications must be followed.

Duties of Headquarters Staff

After all manuscripts have been received in final form from the monograph editor, much of the responsibility for completing the monograph rests with the headquarters office staff. A managing editor from the staff is appointed to supervise production. Thereafter, the following steps are followed:

1. The manuscripts are registered and filed.
2. The material is read and edited critically for style and omissions, ambiguities, or errors that might have been overlooked.
3. The managing editor corresponds directly with the corresponding authors about questions requiring their attention and sends the transfer of copyright form. This procedure may be combined with copyediting and proof stages described below.
4. The manuscripts are typeset, under the direction of headquarters.
5. Proofs for each chapter are sent to the corresponding author for checking. The authors return their proofs to headquarters. The headquarters editorial staff enters all author corrections on the master copies, checking the authors' corrections for style, and returns the master proofs to the typesetter.
6. Each corresponding author receives a set of second proofs of his/her chapter to be used to complete the index by inserting the page numbers. (See section on index preparation for details.)
7. The managing editor and headquarters staff then take appropriate action on the matters outlined below.
   a. Prepares material for the front matter, including a Library of Congress catalog number, ISBN number, and copyright notice.
   b. Arranges with the monograph editor to prepare a suitable preface, checks the alphabetical list of authors and their titles and addresses, compiles the table of contents, and assists the monograph editor in producing the index, which must be completed quickly when proofs are available.
   c. A foreword is prepared for approval by the president(s) of the society(ies) sponsoring the monograph.
   d. Arranges final details for the cover, front matter, index, and other matters.
8. The managing editor compiles the information necessary for obtaining printing cost quotations, discusses them with the vice

president—management and operations, and sends the specifications to several printers for quotations.

9. Quotations received from printers are analyzed critically as to quality, time schedule, and cost, in the order listed, and a printer is selected.
10. The vice president—management and operations and the managing editor, with information made available by the monograph editor and with additional information on printing costs, make a final determination of the number of copies to be printed, unless the number has been stipulated by the board of directors.
11. The printer is informed of the number of copies desired and proceeds with printing and delivering the books.
12. The headquarters staff secures a copyright for the book.
13. After the printing costs and office costs are known, a price per copy is established by the headquarters office staff.
14. All promotion, sales, and distribution are the responsibility of the headquarters office, but suggestions and assistance are welcome from the monograph editor, editorial committee, or any individual.
15. A complimentary copy of the monograph is sent with a letter of appreciation from the headquarters office to each author, to the editor, and to each member of the editorial committee.
16. The corresponding author of each chapter is sent 200 copies of that chapter. If a given chapter has more than one author, the 200 reprints are divided equally or as specified by the authors.

Suggestions for New Monograph Editors

The role of editor for a monograph is a rewarding and worthwhile experience, but requires much time and effort. By the time an editor reads these suggestions, the Society or Societies sponsoring the monograph has been determined by the executive committee and the editorial committee has been appointed in consultation with the editor. How should the editor proceed? The *Publications Handbook and Style Manual*, (1988) Chapter 3, and preceding sections in this manual list the activities associated with the publication of a monograph. What is missing is the sequence of events, items that need emphasis, and sources of assistance. This section is an attempt to provide some of this information.

Past editors can be of assistance, but each monograph is a unique experience. The headquarters staff provides the greatest continuity and experience, and should be consulted from the beginning for assistance and guidance.

The editor should study the list of duties of the editor and editorial committee, as outlined in this manual and the *Publications Handbook and Style Manual* (1988). The selection of the right individuals on the editorial committee determines the ease of the editor's task. It is the responsibility of the editor to be prepared beforehand to provide the leadership in working with the editorial committee and authors. Astute choice of editorial committee members will provide the editor with a cadre of experienced scientists useful in providing peer leadership to authors.

Editors and editorial committees of each monograph establish different working relations on how they share responsibility. The working organization chosen will depend on the interpersonal relations between the editor and the committee. The more people sharing the responsibilities will mean less work for the editor. It may take more time to develop a monograph when more people are involved in the editorial oversight process, unless strict assignments and timetables are accepted. Appendixes A and B in this manual illustrate two models of monograph organization, and modifications of each are common.

Before meeting with the editorial committee, the editor should obtain from the vice president—management and operations at headquarters the guidelines for size and cost of a monograph to know how many pages to devote to a single volume, and the resulting costs. The deliberations of the editorial committee depend on knowing how much can be put into a volume, the number of volumes, and the anticipated audience and total market.

The editor and the editorial committee have the task of targeting the audience, determining the content, and setting priorities of how much text will be devoted to each topic. The tendency of most first efforts is to be all inclusive of subject matter and audience served. Compromises have to be made to determine the important information and the audience that wants and will purchase the information. Part of this discussion includes the fact that contents determine the authors and the availability of authors determine the contents. Some of the activity may be proposed by the feasibility committee, which was appointed to determine the need for the monograph. An approach used in revising an existing monograph is for editorial committee members to review each chapter and propose the magnitude of revision or rewriting required.

The next important step is for the editor and editorial committee to establish the monograph editorial guidelines. It is important to establish as much detail as possible at the beginning of the process. Invariably, some items will not be covered, but it is important to minimize these occurrences. There may be authors from diverse disciplines who will be required to prepare manuscripts in an unfamiliar style. The earlier everyone is familiarized with what is expected of them, the smoother the process. However, do not assume that all authors will read the instructions and guidelines. Repeated and persuasive leadership will be required. Authors who are not members of the society must be provided with a copy of the *Publications Handbook and Style Manual* (1988).

Deadlines must be established, and if a record of a previous monograph timetable is available from headquarters, it will be useful. Set reasonable but not generous deadlines. It will be easier to get people to assume the responsibilities of authorship if deadlines are reasonable, but remember, people tend to work against deadlines. After establishing the deadlines, be prepared for everything to take longer than planned.

Monograph editors should maintain a record of developmental activity for each monograph chapter. See Appendix C in this manual for an example. First, the record will provide an instant summary of progress and identify potential problems as voids in the timetable. Second, the chair of the ASA Monograph Committee will request a quarterly status report using Appendix C for use by headquarters staff and the executive committee to target problems and to coordinate publication schedules.

SSSA BOOK SERIES

The SSSA Book Series follows similar editorial policies and procedures as described for Agronomy Monographs.

BOOKS

The three societies publish hardcover books on an irregular basis. The societies' books, in contrast to their monographs and special publications, and the SSSA series, are not published in a numbered series. Subject matter of books is not usually covered in as much depth as in monographs. On the other hand, a society book covers a broader aspect of a particular subject than does a special publication.

Books can be (i) a compilation of papers presented at special society(ies)-sponsored symposia or conferences normally held at a time other than during the annual meetings; (ii) a compilation of papers written by a number of authors on a particular subject, but not necessarily presented at a symposium; or (iii) in some instances a manuscript written by a single author.

Proposals for books should be submitted to the executive vice president, vice president—management and operations, or the president(s) of one or more societies. The executive committees make the decision whether to publish after consultation with the respective editors-in-chief.

If the papers in the book are to be presented at a special symposium, a separate planning committee is appointed by the executive committee(s) before or at the same time the editorial committee is named. This planning committee is appointed by the sponsoring societies to work closely with the headquarters in developing the meetings. The committee develops the sessions, outlines the topics, and selects the participants who later are authors of the papers published. Manuscripts for books are subject to the normal review procedures and editorial policies of the society(ies) as set forth in this manual and the *Publications Handbook and Style Manual* (1988) and are subject to all policies on style. Approval of a publication does not constitute approval of the manuscript(s) since this approval can come only from the editor or editorial committee of the publication. Duties of the editor, editorial committee, society editors-in-chief, and headquarters staff are similar to those described for monographs.

SPECIAL PUBLICATIONS

Paperbound special publications usually result from timely topics presented in symposia at the annual meetings of the three societies. Each of the three societies has its own special publication series, numbered serially, but ASA, CSSA, and SSSA can also jointly publish in any of the series.

Each of the three societies has guidelines for special publications. In 1974, the executive committees of ASA, CSSA, and SSSA adopted policies relative to special publications that result from symposia presented at the annual meetings. Generally it is preferable not to appoint organizers of the symposium resulting in a special publication to the editorial committee that reviews the publication, although the chair of the organizing committee may be named coeditor of the publication. Under most circumstances none of the authors for the publication will be named as members of the editorial committee.

Organizing Committee

Members who wish to initiate a special publication should submit a proposal to the executive vice president or vice president—management and operations on a special form available from headquarters office. If the proposal is to publish the proceedings of a symposium, it must be submitted at least 26 weeks before the symposium.

The executive committees of the three societies review special publication proposals to determine which of the societies is interested in publishing and financing the publication. The editors-in-chief of the three societies are consulted relative to the appropriateness of having each respective society participate in a proposed special publication. If the publication is jointly sponsored by the three societies the leadership falls to ASA. If the publication is jointly sponsored by two of the societies, such as ASA and CSSA, or ASA and SSSA, the leadership is assumed by CSSA or SSSA.

Editorial Committee

Those submitting a special publication proposal suggest persons who could be asked to serve on the editorial committee. Twice as many persons as is needed for the committee should be suggested. The symposium organizer, if desired, can be suggested for this committee.

It is advisable that at least one member of the editorial committee have experience on the respective society editorial board. Editors-in-chief of the respective societies are ex-officio members of the editorial committee.

Title Page Credits

The organizing committee of the symposium and the editorial committee for the publication will be recognized on the title page of the publication in separate columns.

Deadlines for Symposium Papers

Papers for a special publication, whether invited or volunteered, must be submitted to the organizer of the symposium or the editor of the special publication at least 4 weeks before the date the symposium is scheduled to be presented at the annual meeting.

Handling of Manuscripts

Manuscripts are submitted to the editor, coeditor, or chair of the editorial committee and are assigned to members of the committee for review. Committee members recommend acceptance, revision, or rejection to the chair, who makes the final decision on the acceptability of the manuscripts and communicates this decision to the authors. The chair is also responsible for the thoroughness of review, completeness of coverage, and the scientific quality of the publication.

Although papers in special publications are often the result of an invitation to present work verbally, the printed paper must meet suitable publication standards. If a manuscript does not meet these standards (see "Review of Manuscripts" below), the editor or chair of the editorial committee will require that the author revise it. If the author cannot or does not revise the manuscript to the satisfaction of reviewers in a reasonable time, the manuscript will be omitted from the special publication.

Review of Manuscripts

Manuscripts are reviewed for scientific merit, organization, readability, clarity of presentation, and proper grammar. They must conform to the current requirements of the society's *Publications Handbook and Style Manual* (1988) for style, symbols, and preparation of text.

At least two reviews are obtained for each manuscript. Reviewers must be competent in the subject matter and may or may not be members of the three societies. One of the reviewers should be a member of the editorial committee handling the manuscript. Two favorable reviews are required for acceptance and two unfavorable reviews are required for rejection. Reviewers shall remain anonymous.

Contents of manuscripts that are primarily presentations of original data are reviewed to determine that:

1. Experimental methods and procedures are adequately described;
2. Interpretations and conclusions are valid and consistent with the data presented; and
3. The paper presents significant new information or significant new interpretation of pre-existing data.

Contents of manuscripts that are primarily reviews and interpretations of published data are reviewed to determine that:

1. All aspects of the topic have been adequately covered;
2. Data presented are representative of the information published; and

3. The interpretations and conclusions made are consistent with the data and results presented.

Reviewers have 4 weeks to review the manuscripts and the authors have 4 weeks to make the revisions and return the manuscripts to the editor of the special publication.

When the manuscripts are acceptable for publication, the chair forwards them to the vice president—management and operations at headquarters, who appoints a managing editor to oversee final processing and printing. A design or artwork for the cover of the book may be suggested at the time the manuscripts are submitted or later. Final production procedures are similar to those described for monographs with the exception that the contributor list and subject index are usually not included.

PREPARING A SUBJECT INDEX

A well-prepared subject index can be a valuable tool to a reader, and a poorly prepared index can cause that reader untold frustration.

Preparing an index is a necessary and time-consuming part of book production that must be done quickly, yet requires meticulous attention to detail. Preparation of the index is one of the responsibilities of the book or monograph editor, but headquarters will provide assistance and suggestions.

Four basic methods can be used to prepare an index for a society book:

1. (Most commonly used.) During the review of the manuscript pages, the editor prepares a list of terms that chapter authors follow as they index their chapters. The editor or headquarters staff compile these author-prepared chapter indexes and the editor gives the final review.
2. (Occasionally used.) The editor alone prepares the entire index.
3. (Rarely used.) Authors, without input from the editor, prepare an index for their chapters. The rest of the steps are similar to Method 1. This method is not as satisfactory because there can be a substantial variation among material prepared by individual authors.
4. (Occasionally used.) The editor asks headquarters to hire a professional free-lance indexer to prepare the index with little or no editor input.

There is no definitive answer as to which is better, but indexes prepared by authors and/or editors are likely to be more accurate in content and have better cross referencing of synonymous terms than will an index prepared by a professional indexer. The editor/author, since they are experts on the subject, are also better judges of what should and should not be indexed.

An index prepared by a professional indexer tends to be better technically, the alphabetizing will be more correct, entries are more likely to be indexed by nouns rather than by adjectives, and excess entries will be minimized by judicious use of cross references. Also, an experienced indexer is often better able to keep the needs of the reader in mind.

Since essentially all of the society books have subject indexes prepared according to one of the first three methods listed above, this section gives guidelines for the editor to follow.

Getting Acquainted with Indexes

A subject index is neither a table of contents nor an outline of the text. A table of contents gives an overall view; an index gives specific information. A user looking for calcium requirements of corn, for instance, would look in the table of contents for the chapter on nutritional requirements, but in the index for the term "calcium."

Perhaps the best way to begin is by looking at existing books whose indexes have been particularly useful. Editors should study those indexes critically to determine why they like them, what makes them useful, and to see if they can get ideas about style of entries, cross referencing, and anything else they can include in their own index. In a similar vein, editors should study indexes that have not been useful to them to see what they do not like about them and to avoid making the same mistakes.

If a new book is a later edition of an existing book, editors should consult the index of its earlier editions. The goals would be to retain what is good, and to correct indexing mistakes.

Preparing the Index Outline

A person studying manuscript pages to prepare the index outline chapter authors will follow, should try to put oneself in the place of the reader; continually asking the question, "If I were the reader of this book, what information would I be interested in finding in this index and where would I most likely look for it?" The reader, will look for:

1. Needed information.
2. Information that has been discussed in some substantial detail in text.

The second point is important; it is possible to overload an index with too many minor or even unnecessary items. For example, every crop cultivar, every soil type, every chemical, every person mentioned merely in passing need not be indexed. For example, if corn is mentioned once in a book about wheat, it would not normally be necessary to index corn.

Choosing Index Words

Too few items in an index are as bad as too many items. Broad general categories, each followed by dozens of page numbers, are no help to a reader.

The person preparing an index outline should avoid overly broad categories, such as "water" or "diseases." A word such as "corn" would be too general for a corn monograph index, but perhaps not in another book. Broad categories can be divided into subentries. And subentries, if needed, can be further divided into sub-subentries.

What index words should be included in the outline? Get a clue from such things as chapter titles, subheads, and perhaps sub-subheads and titles of figures and tables. Certainly key words from key sentences should be index items.

Select an average of three to four index words per book page (each book page is equivalent to two manuscript pages). Of course, some introductory pages may produce no index words and other pages may require as many as 10 or more.

Organizing the Index

How does one compile the words and phrases they have selected? A good indexing program on your word processor is one way. Otherwise, index cards can be used; one card for each entry, with indications as to whether it is a main entry, subentry, or sub-subentry. (Be sure to repeat the main entry on each of these cards.) Individual cards make it easy to alphabetize.

Whichever method is used—cards or word processor—begin by grouping key words and phrases into some sort of order; main entries, subentries, sub-subentries. A key phrase such as "nitrogen fertilizer" should be indexed under both words.

Do not forget the value of "see" and "see also" entries, such as "Varieties, see Cultivars," and "Nitrogen, see also Fertilizer."

Make notes regarding particular style items. An editor might decide, for instance, not to use the word "pesticide." If so, an entry such as, "Pesticides—see specific categories such as herbicides, insecticides, fungicides, etc." will be needed.

Think about alphabetizing as entries are prepared. Dozens of entries that begin with "Effect of . . ." are not only difficult to alphabetize, they are also difficult for the reader to use. For example. "Sulfur, effects on crops," is better than "Effects of sulfur on crops."

Try to determine the word the reader would use to find the particular topic. Information on sulfur effects are more likely to be looked for under "Sulfur" than "Effects." A new laboratory technique should not be indexed under the word "New."

Ideally each index word should be a noun. Adjectives, such as "genetic," are not good entries. Nouns used as adjectives can often be converted: "Chicken manure," for example, can be converted to "Manure, chicken." Even when it does not affect alphabetization, an entry such as "Water, infiltration of," is preferable to "Water infiltration" because it is a noun instead of an adjective word form.

The noun should usually be singular; it is easy to recognize redundant entries such as "tree" and "trees" that should be combined. Redundant entries such as "goose" and "geese" are harder to spot once they are alphabetized.

The handling of scientific synonyms first requires some style decisions. Two frequently encountered situations follow:

1. Chemical elements may appear both under their names and their symbols in the text. Use names as main entries in the subject index and select which you prefer for subentries. Whichever form is used, be consistent. Names of chemical compounds should always be written out.
2. Scientific and common names of plants or other organisms can be handled by:
   a. Cross referencing—"*Zea mays* (L.), see Corn," (or vice versa), or,
   b. As subentries under main entries—"Insects (common names)" and "Insects (scientific names)."

Think also of the length of entries. There is a tendency among persons unfamiliar with indexing to make entries too long; particularly sub-subentries, which have the least amount of space available for them. Keep them as short as possible; try for three words or less, never more than six.

Once all terms are compiled and the cards are stacked in alphabetical order (if a word processor is used skip this step), type the entries doublespaced onto 216- by 280-mm (8.5- by 11-inch) paper. The completed outline will look like an index except that it will not have the page numbers after the entries. It will be the responsibility of the editor (or the chapter authors) to supply those page numbers at the time the second page proofs are read.

If the authors will be preparing individual chapter indexes, editors should send their outline to headquarters for duplication and distribution to the authors, along with instructions on how to proceed.

Following the Index Outline

The actual indexing of the chapters is done at the second page proof stage. The authors (or the editor) will refer to the outline as they study the page proofs.

Three steps are involved:

1. Find the key words from the outline in the text,
2. Underline those words with blue pencil, and
3. Write the page numbers into the outline. If several pages are involved, inclusive numbers (156-162) are used.

A blue pencil is suggested so that these underlinings will not be confused with the red proofreading marks. Underlining the words makes it easy to find them later in case page proof corrections force a change in page number.

Authors, most likely, will want to cite words that were not in the entry list. They will be instructed to underline them in the text proof and write them and their page numbers into the outline. The editor will have an opportunity to review these additions.

The authors will send their chapter indexes to headquarters. If the editor has prepared the outline on a word processor, the author indexes will be forwarded to the editor so that the page numbers and additional terms can be added. If the outline was not prepared on a word processor, the chapter indexes will be compiled and page numbers added at headquarters, often with the aid of a professional free-lance indexer who can smooth out rough spots and clear up ambiguities. A typewritten or word-processed copy will be sent to the editor for final review.

When the editor receives the chapter indexes, one of the first tasks will be to analyze each deviation from the outline to see which suggestions to incorporate.

With the final form of the index completed and all page numbers in place, the editor has the first opportunity to see it in its entirety. This is the time to see if major changes should be made. Consider, for instance, if the index is too detailed or not detailed enough.

Four guidelines to keep in mind follow:

1. An unusually large number of entries with only one page number after them is a sign that there are too many categories. Consider if any of these categories can be combined or possibly eliminated.
2. If entries have more than six page numbers after them, this is a clue that the index may not be detailed enough. See if some of these categories can be broken up with one or more subentries.
3. Generally there should not be page numbers after main entries that have subentries after them.
4. A large number of subentries all referring to the same page are not as useful as separate main entries. The main entry "Mineral requirements" may not need subentries if all minerals are listed on the same page. It may be better to have "Mineral requirements" and "Zinc requirements" as separate main entries, rather than having the single entry, "Mineral requirements, zinc."

If other minor or substantial changes are needed, this is the time to make them. Changes are much less costly at this stage of production than after the index has been typeset.

Once the editor is satisfied with the final form of the index, it is returned to headquarters for typesetting.

APPENDIX A

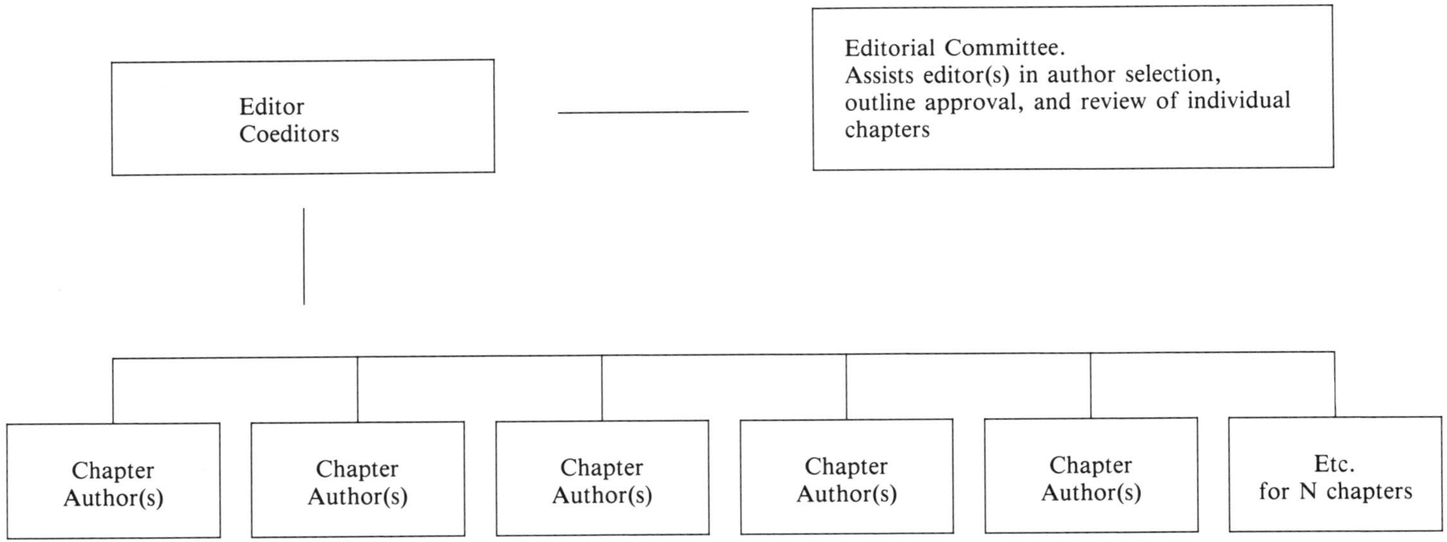

Model I. Editor or coeditors assume entire responsibility for dealing with authors and assuming accountability for progress. Editorial committee only serves in an advisory and peer review role. Editor or coeditors handle all steps of publication preparation without delegation of authority except for chapter review.

APPENDIX B

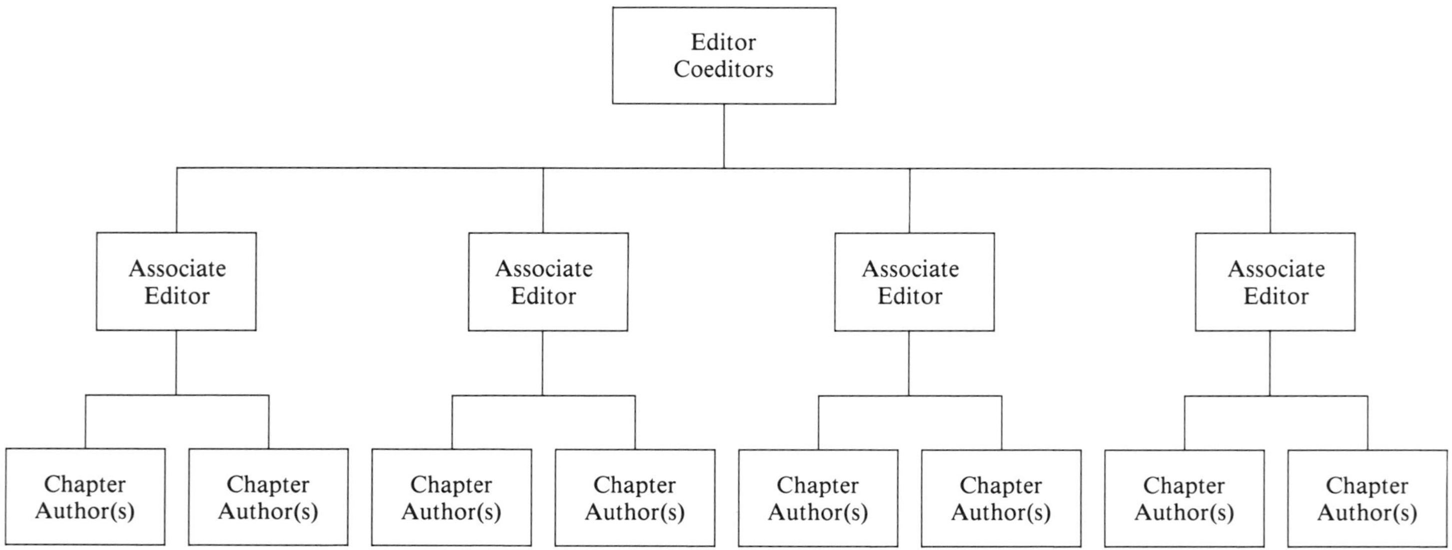

Model II. Editor or coeditors provide overall guidance for publications and are responsible for overall editing of final manuscripts for uniform style and for indexing. Editorial committee members serve as associate editors to handle author correspondence for a subset of chapters within their general areas of expertise. Associate editors assist by assuming responsibility for monitoring progress for a set of individual chapters, for obtaining anonymous peer reviews, and for assuring that the final revision conforms to the reviewer suggestions. In this model, the associate editors are usually senior scientists with previous editorial experience, usually with a journal.

APPENDIX C

ASA-CSSA-SSSA Publication Status Report

Comm. Code No. ______________

Title of Publication ______________ Editor(s) ______________ Date of Report ______________

Chapter											
No.	Short Title	Author(s)	Outline Approved	Literature Review Done	First Draft Rec'd	Peer Reviews to Author(s)	Revision Received	Editor's Review Complete	Final Revision Received	Index Rec'd	Manuscript to HQ
			(date)								
1											
2											
(etc.)											

NOTES

NOTES

NOTES

NOTES

NOTES

NOTES